The Temporal Key: Unlocking the Secrets of Time Travel's Doorway

Hints at a pivotal tool or knowledge for accessing temporal realms

By

Peter Kattan

Book Bound Press

https://web.facebook.com/BookboundPress/

Preface

The allure of time travel has captured the human imagination for centuries, transcending cultures, disciplines, and generations. From ancient myths of gods manipulating the threads of destiny to modern-day scientific inquiries into the fabric of spacetime, the concept of moving through time resonates with a profound, universal yearning: the desire to understand, revisit, and perhaps even reshape the story of existence.

The Temporal Key: Unlocking the Secrets of Time Travel's Doorway embarks on a journey through the intricate tapestry of time, weaving together insights from science, philosophy, history, and culture. This book is not merely an exploration of the mechanics of temporal navigation; it is an invitation to consider time travel as a metaphor for our deepest aspirations, fears, and the boundless potential of human ingenuity.

Time, as a construct, has baffled and inspired thinkers throughout history. In the opening chapters, we delve into diverse perceptions of time—from ancient cyclical views to the linear progression favored by modernity. The book examines the theoretical foundations laid by groundbreaking scientific pioneers like Einstein and the enigmatic implications of quantum mechanics, setting the stage for a thoughtful discussion of whether time travel might move from the realm of fiction to reality.

The journey through these pages is not limited to scientific inquiry. History reveals that the concept of time travel has existed long before the advent of modern science. Myths, legends, and literary narratives across the world have envisioned characters defying the constraints of time. These stories provide rich, imaginative frameworks that reflect humanity's enduring fascination with rewriting the past and glimpsing the future.

As the narrative unfolds, we address the mechanics of temporal navigation, exploring the theoretical devices and energy requirements that make time travel conceivable. Yet, these discussions are tempered by an awareness of the paradoxes and moral quandaries inherent in tampering with timelines. How does one grapple with the butterfly effect? What ethical considerations arise when altering history? These questions lead to deeper reflections on our responsibilities as potential architects of time.

Popular culture has played an indelible role in shaping public perceptions of time travel. From classic literature to blockbuster films and immersive video games, these narratives have not only entertained but also sparked philosophical debates about the nature of existence and the consequences of temporal interference. Through these cultural lenses, we gain a broader understanding of why time travel remains a compelling theme in the human psyche. Beyond the spectacle and the science, this book also contemplates the deeply personal dimensions of time travel. How does our perception of time shape our memories, regrets, and aspirations? Can time travel, whether literal or metaphorical, offer opportunities for personal growth and healing? These questions invite readers to reflect on their own relationship with time and the possibility of second chances—a theme as timeless as the concept itself.

The journey concludes with a forward-looking perspective. Advances in technology, from virtual reality to theoretical physics, hint at tantalizing possibilities for understanding and perhaps traversing time. Yet, the book also emphasizes the importance of the present moment—a lesson as profound as any scientific breakthrough. *The Temporal Key* is both a guide and a companion for those intrigued by the mysteries of time. Whether you approach this book as a curious reader, a science enthusiast, or a seeker of philosophical insights, you will find within these pages a mosaic of ideas to ignite your imagination and challenge your understanding of what it means to exist within the bounds of time. Let us embark on this journey together, unlocking the secrets of time's doorway and embracing the endless possibilities that lie beyond.

Peter I. Kattan   December 2024

Introduction

Time's a wild ride, isn't it? It stitches together moments and memories, shaping who we are. Philosophers scratch their heads over it, artists pour their souls into it, and scientists chase it like it's the Holy Grail. But what if we could peel back the layers of this complex web and discover what's hiding inside? What if time travel isn't just a daydream but something we could actually dive into?

"The Temporal Key: Unlocking the Secrets of Time Travel's Doorway" is your ticket to an adventure that goes beyond the usual nine-to-five grind. We're gonna dig deep into how different cultures and philosophies view time. Think about ancient myths with gods and heroes surfing the rivers of time, mixed with today's mind-blowing physics. It's a mash-up of ideas that have shaped how we see time itself.

Picture this: you're standing at the edge of
time travel. The past, present, and future?
They're all jumbled together in a swirl of
possibilities. We'll break down the science that
makes this dream feel real, from Einstein's
game-changing theories to the crazy
implications of quantum mechanics. What if
reality's more bendy than we've ever thought?
Each new insight gets us closer to figuring out
how to navigate time, pondering the gadgets
that could one day shoot us through time's
hallways.

But hey, with great power comes a boatload of
responsibility. As we dive into the ethical
messiness of time travel, we'll face some tough
questions. What if you could change just one
moment? How would that throw everything off
balance? We'll look at personal growth, too,
exploring how the past can light up our present,

giving us fresh chances and new ways to see our lives.

In our tech-driven world, we'll check out how virtual reality and brain science are changing how we vibe with time. As we sift through the psychological twists of our time perceptions, we'll see how time travel—whether real or just in our heads—affects our hearts and minds. Memory's a big deal; it shapes how we view time and can even help us heal and discover who we really are.

But this isn't just a dry academic stroll; it's a deep dive into what it means to be human. We'll unpack the messy nature of relationships over time, exploring how our ties to past selves and lost loves can change how we think about friendship and intimacy. The way different cultures interpret time adds a colorful layer, showcasing the rich variety of beliefs that shape our understanding of this core part of life.

As we tackle the mind-bending paradoxes of time travel, we'll wrestle with questions that've puzzled thinkers for ages. How do we deal with clashing timelines? What happens when our choices meet destiny? Through this exploration, we'll find wisdom in the journey itself, discovering the importance of soaking up the present while we navigate time's twists and turns.

"The Temporal Key" isn't just a book; it's an invitation to unlock the mysteries of time travel and embark on a transformative journey. Get ready to challenge your views, spark your imagination, and think about the endless possibilities ahead. Are you set to turn the key and step into time's doorway? Adventure's calling, and the only limit is how far you're willing to go into the unknown.

Table of Contents

Chapter 1: The Concept of Time   15
- Different perceptions of time across cultures
- Theories of time in philosophy and science
- Time as a fourth dimension in physics

Chapter 2: The Science of Time Travel   30
- Theoretical foundations of time travel
- Einstein's theory of relativity and its implications
- Quantum mechanics and the multiverse theory

Chapter 3: Historical Perspectives on Time Travel   43
- Time travel in ancient myths and legends
- Notable historical figures who pondered time travel
- Evolution of time travel in literature

Chapter 4: The Mechanics of Temporal Navigation   54
- Theoretical devices for time travel
- Energy requirements and paradoxes
- Navigating timelines and alternate realities

Chapter 5: Time Travel in Popular Culture   68
- Influential films and television shows
- Literature that shaped public perception

- The role of video games in exploring time travel

Chapter 6: The Ethics of Time Travel   73
- Moral dilemmas associated with altering the past
- The butterfly effect and its implications
- Responsibilities of a time traveler

Chapter 7: Time Travel and Personal Growth 79
- Learning from past experiences
- The concept of second chances
- Using time travel as a metaphor for self-improvement

Chapter 8: The Role of Memory in Time Travel   93
- How memory shapes our understanding of time
- The impact of nostalgia and regret
- Techniques for harnessing memory for personal growth

Chapter 9: Time Travel and Technology   98
- Current advancements in technology related to time perception
- Virtual reality as a tool for exploring time
- Future technologies that could make time travel possible

Chapter 10: The Psychology of Time   104
- How humans perceive the passage of time
- Psychological effects of time travel on
individuals
- Coping with time-related anxiety

Chapter 11: The Future of Time Travel   115
- Predictions for time travel advancements
- Potential societal impacts of time travel
- The role of science fiction in shaping future
possibilities

Chapter 12: Interpersonal Relationships Across
Time   121
- Navigating relationships with past selves
- Time travel and its effects on love and
friendship
- Reconnecting with lost connections through
time

Chapter 13: Cultural Interpretations of
Time   134
- How different societies view time and its flow
- The impact of cultural attitudes on time travel
narratives
- Celebrating diverse temporal philosophies

Chapter 14: The Paradoxes of Time Travel   142
- Understanding classic time travel paradoxes
- Resolving conflicts between timelines
- Theories on how paradoxes might be avoided

Chapter 15: Embracing the Journey   153
- Finding meaning in the exploration of time

- The importance of living in the present
- Using time travel as a metaphor for life's journey

- Index: 165

Chapter 1

The Concept of Time

Time is an intriguing concept, isn't it? It's like trying to grasp a handful of water—no matter how hard you try, it always seems to slip away. As you embark on this journey of understanding time, you'll discover that people perceive it in wonderfully diverse ways, each perspective adding a unique layer to our collective experience.

In many Western cultures, time is viewed as a linear progression. Picture a straight road stretching from the past, through the present, and into the future. It's all about the milestones—birthdays, deadlines, and the relentless ticking of the clock. Think of the American work ethic: rising with the sun, tackling tasks with determination, and finally collapsing into bed after a long day, all while keeping a watchful eye on that unforgiving clock. It's a dance of efficiency, productivity, and a constant race against time.

But when you step outside this linear perspective, you enter a world rich with cyclical views of time. Take, for example, the Indigenous peoples of North America. Their understanding of time is like a flowing river, with seasons ebbing and flowing into one another. It's not about rushing forward; it's about being present in the moment, honoring the past, and embracing the future. This cyclical perception fosters a deep sense of connection and continuity, as if life itself is a beautiful, harmonious dance.

Now, let's shift our focus to the realms of philosophy and science, where time takes on even more complexity. Philosophers have

pondered the nature of time for centuries.
Aristotle, for instance, viewed time as a
measure of change—an observer's perception of
events unfolding in sequence. In contrast,
Immanuel Kant argued that time is not an
independent entity; rather, it's a framework
through which we make sense of our
experiences. Imagine trying to appreciate a
piece of art without understanding the artist's
intention—time, in this sense, is the canvas
upon which we paint our lives.

And then there's the scientific perspective.
Ah, the scientists! They've transformed time
into a mathematical enigma. Einstein, with his
wild hair and brilliant mind, introduced us to the
theory of relativity. He revealed that time isn't
as straightforward as we once believed. It bends
and warps, influenced by speed and gravity.
Picture a clock aboard a spaceship racing
through the cosmos; it ticks at a different rate
than a clock resting on Earth. This revelation
can feel dizzying, but it's an essential piece of

the puzzle when we consider the possibility of time travel.

Now, let's delve deeper into the concept of time as the fourth dimension. In physics, we typically think of three dimensions of space: length, width, and height. Time, however, is the dimension that propels everything forward. It's like an invisible thread weaving through the fabric of our universe. When we consider time as a fourth dimension, we begin to see a grander picture. Events are not isolated moments; they're interconnected, forming a vast tapestry woven together by the threads of time.

This perspective opens up exciting possibilities. If time is indeed a dimension, then perhaps it can be navigated, much like we navigate through space. Imagine the potential: if we could find a way to access those temporal realms, we might step back into history or leap

forward into the future. It's a tantalizing
thought, isn't it?

But before we get carried away with visions
of time travel, let's ground ourselves in the
reality of understanding time in all its
complexities. As we explore the various
perceptions of time across cultures, the
philosophical theories that have shaped our
understanding, and the scientific principles that
govern it, we're laying the groundwork for
something truly extraordinary.

Let's take a closer look at these different
perceptions of time. In the Western world, we
often treat time as a commodity. We save it,
spend it, and waste it. The emphasis is on
efficiency and productivity. But in cultures
where time is viewed as cyclical, there's a sense
of harmony and balance. People take their time,
savoring each moment, allowing life to unfold
naturally.

This difference in perception can lead to misunderstandings. Have you ever tried to schedule a meeting with someone from a culture that values flexibility over punctuality? It can be quite the challenge! But when we embrace these differences, we open ourselves to new ways of thinking. We learn to appreciate the beauty of the present moment rather than always racing toward the next deadline.

Now, let's dive deeper into the philosophical side of time. Time has puzzled thinkers for ages. The ancient Greeks had a term for it—"chronos," which refers to chronological time, the ticking clock we all know. However, they also spoke of "kairos," a more qualitative sense of time, the opportune moment. This distinction is crucial. It reminds us that not all moments are created equal. Some moments are ripe for action, while others are meant for reflection.

And then there's the modern debate. Philosophers like Henri Bergson argued that time is more than just a series of moments; it's a flowing experience. He believed that our perception of time is influenced by our consciousness, emotions, and memories. When we think back on a joyous occasion, it feels as if time stretches out, while painful moments seem to drag on forever. This insight reminds us that time isn't merely a measurement; it's deeply personal.

Let's not forget the scientists. The theories of time in physics have transformed our understanding of the universe. Newton saw time as absolute, a constant ticking away the same for everyone, everywhere. But then came Einstein, flipping that idea on its head. He showed us that time is relative, affected by speed and gravity. Imagine standing on a merry-go-round; if you're on the outside, it feels like you're spinning faster than if you're sitting close to the center.

This relativity of time is a game-changer. It suggests that if we could travel at the speed of light, we might experience time differently than those who remain stationary. And that opens the door to all sorts of possibilities—like time travel!

But before we get too carried away, let's ground ourselves in the reality of time as the fourth dimension. When we think of time this way, it's like viewing the universe as a vast landscape. Each moment is a point on that landscape, and we can move through it, exploring different times and places. It's a mind-boggling concept, but it's essential for understanding how we might one day unlock the secrets of time travel.

So, what does all this mean for you? It's an invitation to expand your mind. Consider how your own perception of time shapes your life. Are you always rushing, or do you take time to appreciate the little things? Are you aware of the cyclical nature of life, or do you see it as a straight line?

Reflecting on these questions will help you understand the power of time in your own life. You'll begin to see that time isn't just something that happens to you; it's something you can navigate and shape. And as you embrace this idea, you'll find that the journey of writing your own book becomes an exploration of time itself—a chance to dive into the past, reflect on the present, and envision the future.

Let's take a moment to visualize your journey. Picture yourself sitting at your writing desk, surrounded by the ideas and experiences that have shaped your understanding of time.

Each word you write is a step along the path, guiding you through the past, present, and future. The stories you share, the insights you uncover, and the wisdom you impart will resonate with readers, inviting them to embark on their own journeys of discovery.

As you move forward, remember that every great journey begins with a single step. So take that step, embrace the concept of time, and let it guide you as you unlock the secrets of your own story. The world is waiting for your voice, and the time to share it is now.

In this process, it's important to celebrate your progress, no matter how small. Each sentence you write, each idea you explore, brings you closer to your goal. Embrace the joy of the writing journey. It's not just about the finished product; it's about the growth and transformation that occur along the way.

As you write, consider incorporating your personal experiences with time. How have your perceptions of time evolved? Have you ever felt the weight of time pressing down on you, or have you experienced moments when time seemed to stand still? These anecdotes will create powerful connections with your readers, allowing them to relate to your journey on a deeper level.

And don't forget to invite your readers to reflect on their own relationship with time. Encourage them to consider how their perceptions shape their lives and experiences. This invitation fosters a sense of connection and community, reminding them that they are not alone in their struggles and triumphs.

As you weave together your insights, reflections, and personal stories, remember that

your unique voice and perspective are your
greatest assets. Embrace your individuality, and
let it shine through in your writing. Your
authenticity will resonate with readers, drawing
them in and inspiring them to explore their own
journeys.

As you continue on this path, keep in mind
that the writing process can be filled with
challenges. Writer's block, self-doubt, and
distractions may try to derail your progress. But
remember, every challenge is an opportunity for
growth. Approach these obstacles with a "can-
do" attitude, and remind yourself that you have
the power to overcome them.

When faced with writer's block, try taking a
step back and reflecting on your relationship
with time. What does time mean to you? How
has it influenced your life? Use these reflections
as a springboard for your writing, allowing your
thoughts to flow freely onto the page.

If self-doubt creeps in, remind yourself of your purpose. You have a message to share, and your voice matters. Visualize the impact your words will have on readers' lives. Imagine them finding solace, inspiration, or motivation in your writing. This vision will fuel your determination to keep going.

And when distractions arise, practice mindfulness. Take a moment to breathe, center yourself, and refocus your thoughts. Consider setting aside dedicated writing time, free from interruptions, where you can immerse yourself in the creative process.

As you navigate this journey, remember to celebrate your successes, no matter how small. Each milestone you achieve is a testament to your dedication and hard work. Share your progress with others, and allow their encouragement to lift you higher.

In conclusion, the exploration of time is not just an academic exercise; it's a deeply personal journey. As you dive into the complexities of time, you'll uncover insights that resonate with your own life and experiences. You'll discover that time is not merely a linear progression; it's a rich tapestry of moments, memories, and connections.

Embrace this journey with open arms. Allow it to shape your writing and inspire your readers. As you unlock the secrets of time, you'll find that you're not just telling a story; you're inviting others to join you on a transformative adventure.

So, take that first step. Embrace the concept of time, and let it guide you as you craft your own narrative. The world is waiting for your voice, and the time to share it is now. You have the potential to create something truly remarkable—something that will touch hearts, change minds, and improve lives. Let's embark on this journey together!

Chapter 2

Time Travel: A Wild Ride Through Reality

Time travel—now that's a concept that'll get your brain buzzing! It's like strapping into a rollercoaster that takes you through the very essence of existence. But before we plunge into this cosmic chaos, let's set some groundwork. Think of these basics as the solid beams that keep a barn standing tall. Without 'em, the whole thing could come crashing down around us.

So, what's the deal with time travel? At its core, it's all about bending the rules of time and space. This whole idea isn't just some sci-fi fantasy; it's rooted in the work of some seriously smart folks in physics. They've tossed around ideas that could make your head spin faster than a tornado in a cornfield. And leading the charge? Our pal Albert Einstein.

Einstein's theory of relativity is like the guiding star for anyone wandering through the time travel cosmos. But don't let the fancy lingo throw you off. At its heart, it's pretty simple. It tells us that time isn't this fixed, unchanging thing. Nah, it's more like a river—bending and twisting. Time can speed up or slow down based on how fast you're moving. So, if you're zooming through space, time slows down for you compared to someone just chilling. It's like a cosmic game of tag, where the speedy ones get to bend the rules a bit.

Now, let's chew on what that really means. Picture a spaceship tearing through the universe at light speed. For the folks onboard, they might only experience a few hours, while back on

Earth, years could zoom by. It's a wild thought, right? This opens up all sorts of possibilities. What if we could harness this time-bending magic? What if we could hop into a time machine and witness history firsthand?

But hold your horses; it ain't that easy. There's a catch, and it's a biggie. Enter quantum mechanics—the mysterious, often mind-boggling branch of physics that deals with the tiniest particles in the universe. This is where things get really trippy. Quantum mechanics introduces us to the multiverse theory, which suggests that there are countless parallel universes existing right alongside our own. Every decision we make could create a new timeline, a new version of reality. It's like a choose-your-own-adventure book, but instead of just flipping pages, you're flipping entire worlds.

Imagine this: every time you make a choice—whether it's what to have for breakfast or whether to take that job offer—you're branching out in the tree of time. Some folks say it's like a cosmic buffet—every option you pick leads you down a different path. If time travel is possible, who's to say we wouldn't just hop from one universe to another, exploring all the "what ifs" of our lives?

Now, you might be asking, how do these theories connect to time travel? Buckle up, 'cause here comes the juicy part. If we can manipulate time's flow through relativity and navigate the multiverse through quantum mechanics, we might just find ourselves standing at the edge of time travel. Imagine flipping a switch in a time machine and suddenly being whisked away to ancient Egypt or the Renaissance. The thought alone sends shivers down your spine, doesn't it?

But before we get too lost in the allure of time travel, let's take a sec to appreciate how complex these theories really are. They aren't just doodles on a napkin; they're grounded in some serious scientific research. Einstein's equations, with their intricate dance of numbers and symbols, reveal the underlying structure of our universe. Meanwhile, quantum mechanics, with its strange and often counterintuitive principles, challenges our very understanding of reality.

So, how do we tie these concepts together? How do we weave the threads of relativity and quantum mechanics into a coherent tapestry of time travel? It's all about embracing the unknown. Science, at its core, is about curiosity and exploration. It's about asking questions, pushing boundaries, and daring to dream.

As we dive deeper into the science of time travel, let's remember that every great leap forward starts with a single step. Whether it's a theoretical foundation or a scientific breakthrough, every piece of knowledge brings us closer to unlocking the mysteries of the universe. And who knows? Maybe one day, we'll crack the code and open the door to time travel.

So, as we wrap up this chapter, take a moment to reflect on what we've uncovered. Theoretical foundations of time travel, Einstein's mind-blowing theory of relativity, and the mind-bending possibilities of quantum mechanics. These aren't just abstract concepts; they're the building blocks of a future where time travel could be more than just a dream.

Keep your chin up and your mind open. The journey ahead is filled with wonder and discovery. Embrace the excitement of the unknown, and let your imagination run wild. Because in the grand tapestry of existence, time travel might just be the thread that connects us all. And who knows? The next great discovery could be just around the corner, waiting for you to take that leap of faith.

Now, let's keep moving forward, one step at a time. The adventure is just beginning.

Time travel, in all its complexity, brings up a ton of questions. Like, what happens if you meet your past self? Or what if you accidentally change something in the past? You might just end up with a butterfly effect that flips the entire timeline upside down. It's a bit like that old saying: "A butterfly flaps its wings in Brazil, and it rains in New York." One small change can lead to massive consequences.

Think about it. If you went back to the 1960s and told someone about smartphones, they'd probably think you were nuts. But if you dropped a smartphone into their hands, they'd be in awe. It's mind-boggling to think about how different life would be if you could change even the tiniest detail in the past. Would we have the same technology? The same social norms? It's a wild rabbit hole to go down.

And let's not forget the moral dilemmas that come with time travel. If you could go back and prevent a tragedy, should you? What if your actions saved someone but caused a chain reaction that led to worse outcomes? It's a moral quagmire, for sure. Time travel isn't just a scientific concept; it's a philosophical minefield.

Imagine a scenario: you've got a time machine, and you decide to go back and stop a war. You think you're doing a good thing, but what if that war led to advancements in medicine or technology? You could be messing with fate, and that's a heavy burden to carry.

Plus, there's the whole issue of paradoxes. The grandfather paradox is a classic. If you went back in time and accidentally stopped your grandfather from meeting your grandmother, would you even exist? It's a brain-buster! It raises questions about causality and the nature of time itself.

Let's get real for a second. The idea of time travel is tantalizing, but it's not without its pitfalls. Scientists are still grappling with the implications of these theories. There's so much we don't know, and that uncertainty is part of what makes it all so fascinating.

But here's the thing: science thrives on uncertainty. Every great discovery starts with a question, a curiosity that pushes us to explore the unknown. And who knows what we might find if we keep digging? Maybe one day, time travel will go from fantasy to reality. Maybe we'll unlock the secrets of the universe and discover that time is more flexible than we ever imagined.

And let's not overlook the cultural impact of time travel. Movies, books, and TV shows have played a huge role in shaping our understanding of this concept. Think about classics like "Back to the Future" or "Doctor Who." They've captured our imaginations and made us ponder the possibilities of traveling through time.

These stories often reflect our hopes and fears about the future. They explore the consequences of our actions and the weight of our choices. They remind us that while we might dream of changing the past, we also have to live with the present.

Time travel also brings up questions about identity. If you could go back and change your past, would you still be the same person? Or would those changes alter who you are at your core? It's a deep dive into the essence of self.

And let's not forget the potential for exploration. Imagine being able to witness historical events firsthand. You could see the signing of the Declaration of Independence or walk alongside the dinosaurs. The thrill of experiencing history in real-time is enough to make anyone's heart race.

But, as we've established, it's not all sunshine and rainbows. The risks and ethical dilemmas can't be ignored. If we ever do figure out time travel, we'll need some serious guidelines. Maybe a time travel code of conduct?

In the end, the science of time travel is a wild ride filled with twists and turns. It's a blend of physics, philosophy, and a sprinkle of good old-fashioned imagination. So, as we continue to explore this fascinating topic, let's keep our minds open and our spirits high.

Who knows what the future holds? Maybe one day, we'll find ourselves standing in a time machine, ready to take that leap into the unknown. And when that day comes, we'll be ready to embrace the adventure of a lifetime.

So, keep dreaming big. The universe is vast, and the possibilities are endless. Time travel might just be the next great frontier, waiting for us to explore. And as we journey forward, let's remember that every step we take brings us closer to unlocking the mysteries of existence.

You got this! Let's keep pushing the boundaries of what we know, one curious question at a time. The adventure is just getting started.

Chapter 3

Historical Perspectives on Time Travel

T

ime travel isn't just some wild idea that popped up in the latest sci-fi flicks. Nah, it's been around for ages, stitched into the fabric of our oldest tales, echoing through time like an old song you can't shake off. From ancient myths to the thoughts of some seriously smart folks, the idea of bending time has fueled our imaginations and dreams. So, let's take a little trip down memory lane, yeah? We'll dig into how our ancestors saw time travel, who the big thinkers were that dared to tackle its mysteries, and how literature has morphed to embrace this mind-bending concept.

First off, let's jump into the treasure chest of ancient myths and legends. Across cultures, time travel pops up like a recurring theme, tying humanity together. Ever heard of Orpheus in Greek mythology? This dude ventured into the underworld to snag his beloved Eurydice back. His journey wasn't just about hopping realms; it was a heartfelt exploration of love and loss that defied time itself. And then there's the Mahabharata from Hindu lore, where King Revaita travels to meet Brahma. When he returns, he finds that ages have passed on Earth! These tales remind us that time isn't a straight shot; it's more like a winding river, full of twists and turns.

Now, let's shine a light on the brilliant minds from history who chewed on the nature of time. Take Sir Isaac Newton, for instance. He viewed time as absolute—like a strict teacher keeping order in the universe. But then came along the likes of H.G. Wells, who flipped the script with his classic, "The Time Machine." This wasn't just a story; it was a game-changer that challenged the norm. Wells showed us that time could be navigated, like a ship sailing through uncharted waters. And how could we forget Albert Einstein? His theories of relativity didn't just shake up physics; they cracked open the door to time travel possibilities. Einstein's brain danced with the idea that time could stretch and bend, like a rubber band, leading us to question everything we thought we knew.

As we wander deeper into this historical landscape, let's take a moment to appreciate how literature has evolved around the idea of time travel. It's a fascinating journey, kinda like watching a caterpillar turn into a butterfly. Back in the day, time travel was often seen as a fanciful notion, tucked away in fairytales and folklore. But as society progressed, so did our understanding of time. The 20th century rolled in with a wave of creative works that delved into the complexities of temporal journeys. Authors like Ray Bradbury and Kurt Vonnegut tackled time travel with urgency and depth, weaving it into social commentary and philosophical musings. In "A Sound of

Thunder," Bradbury warns us about the butterfly effect, showing how a tiny change in the past can send ripples through time with catastrophic results. Vonnegut's "Slaughterhouse-Five" takes us on a wild ride through time, blurring the lines between past, present, and future, forcing us to confront the absurdity of war.

So, what's all this mean for you, dear reader? It means that time travel isn't just some whimsical daydream; it's a mirror reflecting our deepest desires and fears. It speaks to our longing to figure out our place in the universe and to make sense of the chaos swirling around us. By diving into these historical perspectives, you're not just learning about time travel; you're tapping into a rich tapestry of human experience that can light a fire under your own writing journey.

When you sit down to jot down your thoughts on time travel, remember to draw inspiration from those ancient myths and legends. Let them fuel your creativity and remind you that stories have the power to transcend time. Embrace the spirit of those notable figures who dared to question and explore. Your unique voice and perspective are your greatest assets, and they'll shine through in your writing.

Now, here's a little exercise for you. Take a moment to reflect on your own experiences with time. Ever felt like you were stuck in the past? Or wished for a do-over to rewrite a moment in your life? Jot down your thoughts and feelings. These personal anecdotes will not only enrich your writing but also create a powerful connection with your readers. They'll see themselves in your words, and that's where the magic happens.

Remember, this journey of writing about time travel is as much about discovery as it is about sharing knowledge. Each chapter you craft is a step forward, unlocking new insights and perspectives. So, embrace the process. Celebrate your small wins, whether it's finishing a paragraph or finding the perfect metaphor. Each victory brings you closer to your goal, and I believe in your potential to create something truly transformative.

As you weave together the threads of ancient myths, historical figures, and literary evolution, keep your heart open and your mind curious. The world of time travel is vast and filled with possibilities. You have the key to unlock its secrets, and I can't wait to see where your journey takes you. So, grab that pen, let your imagination soar, and remember: the past, present, and future are waiting for your unique voice to tell their story. You can do this!

Now, let's get into some of the nitty-gritty. What about the science behind time travel? You might be thinking, "Isn't that just for nerds?" But hang on a sec. The science of time travel isn't just for the geek squad; it's fascinating stuff that's got everyone buzzing. Theoretical physicists have been kicking around ideas about wormholes and time dilation. Wormholes, for instance, are like shortcuts through space-time. Imagine folding a piece of paper and poking a hole through it; you could jump from one side to the other in no time. Sounds like magic, right? But it's all grounded in the mind-bending theories of Einstein.

And what about time dilation? Ever been on a long road trip and felt like time was crawling? Well, according to Einstein's theory of relativity, if you were to travel at near-light speed, time would actually slow down for you compared to folks back on Earth. So, in a way, you could be time traveling! It's wild to think about, but it's all based on solid science.

Now, let's pivot a bit and chat about how time travel is portrayed in pop culture. Movies, TV shows, and books have taken this concept and run with it, giving us a treasure trove of stories. Think about "Back to the Future." Marty McFly jumps in a DeLorean and zooms back to the '50s. It's a fun flick, but it also dives into the butterfly effect, showing how one little change can mess everything up. Then there's "Doctor Who," where the Doctor zips through time and space in a blue police box. This show has become a cultural phenomenon, mixing adventure, humor, and some serious moral dilemmas.

And let's not forget about the Marvel Cinematic Universe. They've taken time travel to a whole new level with movies like "Avengers: Endgame." The heroes use time travel to undo the chaos caused by Thanos. It's a thrilling ride, but it also raises questions about destiny, free will, and the consequences of our actions.

So, why do we love time travel stories so much? Maybe it's because they allow us to escape reality for a bit. We can explore different eras, meet historical figures, or even right past wrongs. It's like a cosmic do-over. Plus, time travel stories often come with a hefty dose of philosophical questions. What if you could change your past? Would you? And what would that mean for your future? It's a mind-boggling thought that gets us all pondering our choices.

Now, let's take a moment to look at how time travel connects to our personal lives. We all have moments we wish we could revisit or change. Maybe it's that awkward conversation you had or a chance you didn't take. Time travel in stories lets us explore those feelings without the actual consequences. It's a safe space to reflect on our regrets and dreams.

So, as you sit down to write about time travel, think about your own experiences. What moments would you revisit if you could? How would you change them? This personal touch can make your writing resonate with readers. They'll see a piece of themselves in your story, and that connection is powerful.

Let's not forget the role of technology in shaping our views on time travel. With advances in science and technology, it feels like we're inching closer to making the impossible possible. Quantum physics, artificial intelligence, and even virtual reality are all playing a part in how we think about time. Who knows? Maybe one day we'll figure out how to bend time to our will.

But with great power comes great responsibility, right? If we ever do crack the code of time travel, we'll need to think long and hard about the ethical implications. Should we change the past? What about the potential to exploit time travel for personal gain? These questions can spark some deep discussions and are worth considering as you write.

So, let's wrap this up. Time travel is a concept that's been around for ages, woven into our myths, explored by brilliant minds, and brought to life in literature and pop culture. It's a reflection of our desires, fears, and the human experience. As you embark on your own writing journey, draw inspiration from the past, reflect on your own experiences, and embrace the endless possibilities of time travel.

You've got a unique voice, and the world is waiting to hear it. So grab that pen, unleash your imagination, and let your stories unfold. The past, present, and future are all yours to explore. Happy writing!

Chapter 4

The Mechanics of Temporal Navigation

Alright, folks, let's get into the nitty-gritty of time travel. We're not just talking about jumping into a DeLorean or twisting some TARDIS knobs. Nah, we're diving into the real deal—the devices, the energy, and how to navigate the wild waters of timelines and alternate realities. So, strap in, 'cause this is gonna be a wild ride!

Let's kick things off with the devices we might use for time travel. Picture this: a machine that opens a door to the past or future. Sounds like a plot twist from a sci-fi flick, right? But there's a ton of theories out there! Wormholes, time machines, you name it. The options are as vast as the cosmos. You've

probably heard of the Einstein-Rosen bridge—
it's like a shortcut through space-time. Imagine
taking a back road instead of the usual route, but
this back road could drop you off in a different
era. The idea is simple: if you could whip up a
stable wormhole, you might just step through it
and land in a totally different time.

But wait a sec! Crafting such a gadget isn't
as easy as flipping a switch. We're talking about
harnessing some serious energy here. That leads
us to the next hot topic: energy requirements
and those pesky paradoxes. Picture this:
powering a time machine would take energy
levels that make a nuclear reactor look like a
candle. Some brainy scientists even talk about
"negative energy." Sounds a bit creepy, huh?
But it's all about bending the laws of physics.

Now, let's not gloss over the paradoxes that
come with time travel. You've heard of the
grandfather paradox, right? If you went back in

time and accidentally stopped your grandparents
from meeting, what happens to you? It's a real
head-scratcher! Some people think that stepping
into the past could create alternate realities. You
might leave your original timeline behind and
kickstart a brand new one. It's like tossing a
pebble into a pond—each ripple stands for a
different outcome.

Navigating through timelines and alternate
realities is like being a pilot in a cosmic storm.
You gotta know your destination and steer clear
of the trouble spots. Imagine you're trying to hit
a specific moment in time. You've gotta plot
your course like a pro, considering all the
variables. What if you land in a timeline where
dinosaurs are still stomping around? Or one
where the internet never got off the ground? The
possibilities are endless, but so are the risks.

Think of your timeline like a tree. Each
branch is a different decision, a different path.

When you make a choice, you're growing a new branch. But what if you wanna go back and change something? You might find yourself on a completely different path, one you never meant to take. That's the beauty and the headache of navigating through time.

So, how do we wrap our heads around all this? It's about understanding the mechanics at play. First, get cozy with the theoretical devices. Dive into those wormholes, black holes, and whatever else the science whizzes have come up with. Next, ponder the energy needs. What would it take to power your journey? Finally, think about the paradoxes and alternate realities. What might you bump into along the way?

As you explore these ideas, remember that mastering time travel hinges on your curiosity and creativity. Embrace the unknown, and don't shy away from asking questions. Every genius inventor started with a wild idea and a burning desire to explore.

This ain't just about science; it's about imagination. Your brain is a powerful tool, and with it, you can unlock the secrets of temporal navigation. So grab your metaphorical compass and get ready to chart your course through the fabric of time. You've got this!

Now, let's break it down into some actionable steps.

Research Theoretical Devices: Dig into the theories surrounding time travel. Look up wormholes, time machines, and other concepts. Get a feel for what's out there.

Explore Energy Requirements: Think about the energy needed for your journey. What are

the implications of using negative energy? How
would you harness it?

Understand Paradoxes: Get familiar with
classic time travel paradoxes. Reflect on the
implications of changing the past and how that
might affect your present.

Visualize Timelines: Draw out your own
timeline. Map out decisions and their potential
outcomes. See how choices can branch off into
alternate realities.

Stay Curious: Keep asking questions. The
more you explore, the more you'll understand
the complexities of time travel.

By breaking it down into these steps, you'll
not only grasp the mechanics of temporal

navigation but also cultivate a mindset that embraces exploration and discovery.

Now, as you venture forth into this captivating world, remember that every challenge is just a stepping stone to growth. You're on a journey, and it's just as important as the destination. So keep that enthusiasm alive, and let your imagination soar! The universe is waiting for you to unlock its secrets.

Let's dig deeper into each of these points, shall we?

When it comes to theoretical devices, the imagination runs wild. There's a ton of speculation on how we could actually time travel. Some folks are all about wormholes— those theoretical passages that could connect distant points in space and time. Think of it like a cosmic tunnel. You could enter one end and

zip out the other in a different era. But creating
one? That's a whole other ball game. It's not
just about theory; it's about physics, energy, and
a whole lot of "what ifs."

Then there's the concept of time machines.
We've all seen movies where a gadget zaps
someone back to the past or forward to the
future. But what would that look like in reality?
Could it be a sleek, shiny device, or maybe
something more like a modified phone booth?
Who knows! The point is, while we can dream
about these machines, the reality is a bit more
complicated. We're still scratching our heads
over the actual mechanics of making it happen.

And let's not forget the energy aspect. To
power a time machine, we'd need a ridiculous
amount of energy. Some scientists have
suggested harnessing negative energy—
basically, energy that's the opposite of what
we're used to. It's a wild concept, but if it

works, it could open up new avenues for time travel. But here's the kicker: we don't even fully understand what negative energy is, let alone how to create it. So, we're left in this intriguing limbo of possibilities.

Now, the paradoxes? Oh boy, they're a doozy. The grandfather paradox is just the tip of the iceberg. Imagine going back and accidentally stopping your own birth. What happens then? Does your existence just fade away? Or do you create a new timeline where you never existed? It's mind-boggling. Some folks argue that every time you change something in the past, you create a new branch in the timeline. So, maybe you don't erase your existence; you just create a version of reality where you don't exist. It's like a cosmic game of "what if."

Navigating these timelines is a whole other kettle of fish. You gotta be sharp, like a pilot flying through a storm. Each decision you make can lead you down a different path. If you think about it, every choice we make creates a ripple effect. It's like throwing a stone into a pond—those ripples represent the countless possibilities of where you could end up.

And let's be real here—sometimes, we don't even know what we're getting into. What if you land in a timeline where the dinosaurs never went extinct? Or one where the internet is just a concept? You could be in for a wild ride, and not always in a good way. You could end up in a reality where everyone speaks a language you don't understand or where technology is centuries behind.

So, how do you prepare for this kind of adventure? Well, it starts with visualization. Picture your timeline as a tree, with branches representing your choices. Each decision grows a new branch, and if you decide to change course, you might find yourself on a totally different path. It's a visual way to understand the complexities of time travel and how your choices shape your reality.

As you think about all this, stay curious. Ask questions. Don't just take things at face value. The more you dig into these concepts, the clearer they'll become. Remember, every great inventor started with a wild idea and a desire to explore.

In the end, this isn't just about science; it's about creativity. Your imagination is a powerful tool, and with it, you can unlock the mysteries of time travel. So, grab that metaphorical compass, chart your course, and get ready to dive into the unknown.

And let's not forget the importance of having fun along the way! Time travel is a thrilling concept, full of adventure and possibility. Sure, it can get a bit complicated, but that's part of the excitement. Embrace the challenges and enjoy the journey.

Now, let's break it down into a few more actionable steps.

1. **Keep Learning**: Time travel theories are always evolving. Stay updated on new research and ideas. The more you know, the better prepared you'll be for your journey.

2. **Experiment with Visualization**: Try drawing your own timelines. It's a fun way to see how choices can lead to different outcomes. You might even discover new paths you hadn't considered before.

3. **Discuss with Others**: Talk to friends or join online forums about time travel. Sharing ideas can spark new thoughts and help you see things from different perspectives.

4. **Embrace the Unknown**: Don't be afraid to explore uncharted territory. The best discoveries often come from stepping outside your comfort zone.

5. **Reflect on Your Choices**: Think about how your decisions shape your life. It's a great way to understand the ripple effect of your actions.

By breaking it down into these steps, you'll not only grasp the mechanics of temporal navigation but also cultivate a mindset that embraces exploration and discovery.

So, as you set out on this fascinating journey, keep in mind that every challenge is just a stepping stone to growth. You're on a quest, and it's just as important as the destination. Keep that enthusiasm alive, and let your imagination take flight! The universe is waiting for you to unlock its secrets.

And who knows? Maybe one day, you'll be the one to crack the code of time travel. Now that's a legacy worth leaving behind!

Chapter 5

Time Travel in Popular Culture

Time travel, huh? It's a wild concept, one
that tickles our imaginations and pulls us into
stories that dance through the ages. From the
flickering lights of a movie screen to the pages
of a gripping novel, time travel has woven itself
into the very fabric of our culture. Let's take a
stroll through this fascinating realm, exploring
how films, literature, and video games have
shaped our understanding and perception of
time travel.

Picture this: the moment you sit down to
watch a classic film like "Back to the Future."
That DeLorean, with its sleek lines and sci-fi
gadgets, revs up your excitement. You're not

just watching Marty McFly and Doc Brown; you're diving headfirst into a world where the past and future collide. This film, along with others like "The Terminator" and "Groundhog Day," doesn't just entertain—it plants seeds of curiosity about time travel in our minds. These stories challenge us to think about the choices we make and the ripple effects they create.

Then there's television. Oh boy, where do I even start? Shows like "Doctor Who" have captivated audiences for decades, blending adventure with philosophical musings on time. The Doctor, with their quirky personality and timey-wimey technology, invites us to ponder the possibilities of time travel. Each episode is a new adventure, a new lesson, and a reminder that time isn't just a straight line; it's a vast, swirling galaxy of potential. And let's not forget "The Flash," where speed and time travel intertwine, giving us a fresh perspective on heroism and sacrifice.

Now, let's flip the page to literature. Books like H.G. Wells' "The Time Machine" didn't just kickstart the genre; they sparked conversations about society, progress, and the human condition. Wells' work opened the door to exploring the moral implications of time travel. What happens when we meddle with time? Can we truly change the past, or are we just setting ourselves up for a cosmic joke? These questions linger in our minds long after we've closed the book.

But wait, there's more! Video games have also stepped into the time travel arena, and boy, have they made their mark. Titles like "Chrono Trigger" and "Braid" invite players to manipulate time in ways that are interactive and engaging. You're not just a passive observer; you're the one making choices that ripple through time. These games allow us to experience the consequences of our actions firsthand, making the concept of time travel not just a thought experiment but a tangible experience.

So, what's the takeaway from all this? It's simple: time travel in popular culture is more than just a plot device. It's a mirror reflecting our hopes, fears, and dreams. Each film, book, and game invites us to explore our relationship with time, urging us to consider how our choices shape our destinies. It's a powerful reminder that, while we may not have a DeLorean or a TARDIS at our disposal, we all have the ability to reflect on our past and envision our future.

As we wrap up this exploration, remember this: the stories we tell about time travel are not just about the mechanics of moving through time. They're about understanding ourselves, our choices, and the beautiful, messy journey of life. So, whether you're watching a movie, reading a book, or playing a game, take a moment to appreciate the magic of time travel. Embrace the lessons it offers, and let it inspire you to make the most of your own time.

Now, let's keep moving forward. There's more to discover on this journey, and I can't wait to see where it takes us next!

Chapter 6

The Ethics of Time Travel

Time travel—what a wild ride, huh? The thought of zipping back to the past or leaping into the future stirs the imagination like a pot of gumbo simmering on the stove. But hold on just a second! Before you grab your temporal keys and dash off to change history, let's take a good look at the ethics of it all. This ain't just a sci-fi movie; it's real-life stuff we gotta think about. So, let's dive into the moral dilemmas, the butterfly effect, and the responsibilities that come with being a time traveler.

Imagine you've got the chance to alter a moment in time. Maybe it's to save a loved one from a tragic fate or to prevent a colossal

mistake that changed the course of history.
Sounds tempting, right? But wait! What about
the ripple effects of that change? You see, every
action has consequences. The moral dilemmas
of altering the past can weigh heavy on the
heart. It's like standing at a crossroads, where
every choice branches out into a myriad of
outcomes. Do you play God? Are you prepared
to shoulder the burden of your decisions?

Let's talk about the butterfly effect. You've
probably heard the saying, "A butterfly flaps its
wings in Brazil and causes a tornado in Texas."
It's a vivid way to illustrate how small actions
can lead to monumental consequences. In the
realm of time travel, it's not just about the
immediate changes you make. It's about the
long-term implications that can spiral out of
control. You might think you're just changing
one little thing, but who knows how that'll
affect everything else?

Think about it this way: you decide to warn your younger self about a poor choice. You think you're doing them a favor, but that warning could set off a chain reaction that alters their entire life trajectory. Maybe they don't meet the person who would've become their best friend. Or perhaps they miss out on an opportunity that would've led them to a fulfilling career. It's a delicate dance, my friend, and one misstep can lead to a waltz of chaos.

Now, let's get to the heart of the matter—the responsibilities of a time traveler. If you're lucky enough to hold the keys to the past and future, you've got a duty to wield that power wisely. You can't just go hopping through time like a kid in a candy store. It's a serious business, and you gotta approach it with a sense of respect and responsibility.

First off, think about the people affected by your actions. Each person you encounter has their own dreams, struggles, and paths. Altering their lives without their consent is like stepping into someone's home and rearranging their furniture without asking. It's not just rude; it's downright dangerous. You've got to consider the impact on their lives and the lives of those around them.

Moreover, there's the question of intention. Why are you traveling back in time? Is it to right a wrong or simply for personal gain? If your motives are selfish, you might just end up causing more harm than good. It's crucial to reflect on your reasons and ensure they come from a place of genuine care and empathy.

And let's not forget about the wisdom that comes with age. As you navigate through time, you'll likely gather insights and experiences that can shape your understanding of the world. Use that knowledge to guide your decisions. Just because you can do something doesn't mean you should. The power of choice is a double-edged sword, and it's up to you to wield it wisely.

So, how do you balance the thrill of time travel with the weight of ethical responsibility? Start by asking yourself some tough questions before you embark on your journey. What are the potential consequences of your actions? How will your choices affect the lives of others? What lessons can you learn from the past that will guide your present and future decisions?

It's all about mindfulness, folks. Take a moment to breathe, reflect, and consider the broader picture. The more you think about the ramifications of your actions, the better equipped you'll be to navigate the complexities of time travel.

In conclusion, time travel isn't just about zipping through the ages; it's about understanding the moral implications of your journey. The butterfly effect teaches us that even the smallest actions can have far-reaching consequences. And with great power comes great responsibility. So, if you find yourself holding the temporal keys, use them wisely. Your choices matter, and the impact you have on the lives of others is profound. Embrace the journey with a heart full of empathy and a mind open to the lessons of time. You've got this!

Chapter 7

Time Travel and Personal Growth

Time, huh? It's a slippery little thing. Like a river that never stops flowin', just keeps on movin'. But what if I told ya we could dip our toes back into that river? What if we could learn from the ripples of our past? That's the magic of time travel—not the sci-fi kind, but the one that happens in our minds and spirits. It's not just about zippin' through the ages; it's about takin' a good, hard look at where we've been and how we can grow from it.

Learning from our past experiences is like diggin' through the wreckage of a ship and findin' treasure. We've all sailed through storms, faced waves that felt like they'd swallow us whole, and yeah, sometimes we've capsized. But every experience, whether it's a high or a low, is a lesson wrapped in a memory. Think about it: every time you stumbled, every time you soared, there's wisdom in that. It's like an old buddy whisperin' in your ear, "Hey, don't do that again!" or "That felt good, let's do it more!"

Now, picture this: you're chillin' on your porch, sippin' sweet tea, and reflectin' on your life. What do you see? Moments of joy, sure, but also those times when you thought you were down for the count. Maybe you lost a job or a relationship fell apart. But here's the kicker—those experiences shaped you. They made you tougher. They taught you what you want and what you don't want.

Let's talk about second chances. Ain't it a
beautiful thought? The idea that life might give
you another shot at somethin' you thought was
lost? It's like findin' a forgotten dollar bill in
your winter coat pocket. You thought you were
broke, but surprise! There's a little extra to
spend.

Second chances remind us that the past
doesn't have to define our future. You can step
back into the ring, ready to fight again. Maybe
you didn't ace that job interview last time, but
this time? You're wiser. You've learned from
the last go-around. You know what to say, how
to present yourself. You've traveled back in
time in your mind, armed with knowledge and
experience.

Using time travel as a metaphor for self-
improvement is like havin' a magic wand that

lets you rewrite your story. You're not literally jumpin' into a DeLorean, but you're certainly reflectin' on your past choices and thinkin' about how they've shaped you. It's about recognizing that every moment is a chance to evolve.

Imagine you're a character in your own story. You're not just a passive observer; you're the hero. And heroes grow. They learn from their missteps, they embrace their failures, and they use those lessons to forge ahead. Every time you think about a past mistake, you're givin' yourself the opportunity to learn and adapt.

So, let's break it down a bit. Here's how you can use these concepts to fuel your personal growth:

First off, reflect on your past. Set aside some time to think about your life. What were the pivotal moments? What lessons did they teach you? Jot down your thoughts. This ain't just a stroll down memory lane; it's an opportunity to gather wisdom.

Next, embrace those second chances. If there's somethin' you want to revisit—be it a relationship, a career path, or a personal goal—don't shy away from it. Approach it with the mindset of growth. You've got the tools now; use 'em!

Now, visualize your future. Picture where you want to be. What steps do you need to take to get there? This is your time travel, your mental journey to a better you. The future is bright, and you're the one holdin' the compass.

Then, create a growth plan. Take what you've learned and map it out. Set goals that reflect your newfound wisdom. Break 'em down into bite-sized pieces so you can tackle 'em one at a time.

And don't forget to celebrate your wins. No matter how small, every step forward is a victory. Acknowledge your progress and give yourself credit for the hard work you've put in. You're on a journey, and every mile counts.

Remember, time travel in the metaphorical sense is about perspective. It's about lookin' back to see how far you've come and using that knowledge to propel yourself forward. Life's a series of lessons, and each one adds a new layer to who you are. So embrace your journey, learn from the past, and don't be afraid to take those second chances.

You've got this, my friend. The door to your personal growth is wide open, and the key? It's in your hands. So step through that doorway, and let's see where this journey takes you. Your future self is waitin' on the other side, and I promise you, they're proud of the work you're doin' today.

Let's dive deeper into this whole time travel gig. Imagine if you could literally go back in time. What would you change? Maybe you'd avoid that cringe-worthy moment at the party where you tripped and spilled your drink. Or perhaps you'd take a different path in your career, choosing a job that didn't make you feel like you were stuck in a hamster wheel. But here's the catch: every single choice, every misstep, every awkward moment—those are the building blocks of who you are today. You might think you'd rewrite history, but what if that history is what makes you uniquely you?

And you know what? The past isn't just a series of mistakes. It's also filled with triumphs. Remember that time you aced a presentation or finally finished that project you'd been procrastinating on? Those moments are gold. They remind you of your capabilities and your strength. They're like little badges of honor you can carry with you.

So, how do we get better at this whole reflection thing? Well, journaling can be a game changer. Seriously, grab a notebook or open up a document on your computer. Write down your thoughts, your feelings, your experiences. It doesn't have to be fancy. Just let it flow. You'll be surprised at what comes out when you give yourself the space to think.

And let's not forget about the power of storytelling. Sharing your experiences with others can be incredibly cathartic. Whether it's over coffee with a friend or on a blog for the world to see, talkin' about your journey helps you process it. You might even inspire someone else to take their own leap of faith.

Now, let's get back to second chances. They're not just a nice thought; they're a real opportunity for growth. Think about that time you tried to learn a new skill and failed miserably. Maybe you picked up a guitar and couldn't even strum a chord. But what if you picked it up again? With practice, you'd get better. The same goes for life. You can try again. You can take those lessons and apply them.

Ever hear of the saying, "What doesn't kill you makes you stronger"? It's true! Every setback can be a setup for a comeback. It's all about perspective. Instead of lookin' at failures as dead ends, see 'em as detours. They're just part of the journey, not the destination.

And visualization? Man, that's powerful stuff. Picture your future self. What do they look like? What have they accomplished? How do they feel? Visualization isn't just daydreaming; it's about setting your intentions. When you can see where you want to go, you're more likely to take the steps to get there. It's like having a roadmap for your life.

Creating a growth plan is where the rubber meets the road. You've reflected, you've visualized, now it's time to act. Break your goals down into manageable steps. Want to run a marathon? Start with a mile. Want to write a book? Write a page. Small steps lead to big changes. And remember, it's okay to stumble along the way.

And hey, don't forget to celebrate! Every little win matters. Did you go for a run when

you didn't feel like it? Celebrate! Did you finish
that project at work? Celebrate! Life's too short
to skip over the good stuff. Acknowledge your
progress, no matter how small. It keeps you
motivated and reminds you that you're on the
right track.

So, what's the takeaway here? Time travel,
in this sense, is about perspective and growth.
It's about looking back to see how far you've
come and using that knowledge to propel
yourself forward. Life's a series of lessons, and
each one adds a new layer to who you are.
Embrace your journey, learn from the past, and
don't be afraid to take those second chances.

You've got this, my friend. The door to your
personal growth is wide open, and the key? It's
in your hands. So step through that doorway,
and let's see where this journey takes you. Your
future self is waitin' on the other side, and I

promise you, they're proud of the work you're
doin' today.

Now, let's get a bit more personal. Think
about a time when you felt lost. Maybe you
were in a dead-end job, or a relationship that
just wasn't working. It's a tough place to be,
right? But here's the thing: those moments of
uncertainty can lead to the most profound
growth. They force you to reassess, to dig deep,
and figure out what you really want.

Take a moment to think about it. What did
you learn from that time? What would you do
differently now? Those reflections are like gold
nuggets, just waitin' to be discovered. They're
the insights that'll guide you as you move
forward.

And what about fear? Fear can be a major
roadblock. It can keep you stuck in your

comfort zone, afraid to take that leap. But here's the kicker: most of the time, what we fear isn't nearly as scary as we make it out to be. It's like that monster under the bed—it seems terrifying in the dark, but when you turn on the light, it's just a shadow.

So, how do you tackle fear? Start small. Take tiny steps outside your comfort zone. Maybe it's speaking up in a meeting or trying a new hobby. Each small victory builds your confidence, and before you know it, you're tackling bigger challenges.

And let's not forget about the people in your life. Surround yourself with folks who lift you up, who inspire you to be better. Those connections can be a huge source of support and encouragement. Share your goals, your fears, your dreams. You'd be amazed at how much strength you can draw from others.

In the end, it all comes down to this: personal growth is a journey, not a destination. It's messy, it's complicated, and it's often uncomfortable. But it's also beautiful and transformative. So embrace the chaos, learn from your past, and keep moving forward.

You've got this, my friend. The future is bright, and it's all yours for the takin'. So go ahead, step into that river of time, and let it guide you to where you're meant to be. Your future self is waiting, and they're ready to celebrate all the amazing things you're gonna accomplish.

Chapter 8

The Role of Memory in Time Travel

Memory is a curious thing, ain't it? It's like a double-edged sword, shaping how we perceive time and our place in it. Think about it for a moment. When you reminisce about the past, those memories aren't just snapshots—they're like portals, transporting you back to moments that have defined you. Memory has a way of coloring our understanding of time, bending it, stretching it, and sometimes even distorting it. It's through memory that we weave the tapestry of our lives, stitching together experiences, emotions, and lessons learned.

But hold on, let's not rush. Memory is also where nostalgia and regret come into play. Ah, nostalgia—it's that sweet ache in your heart when you think back to simpler times, isn't it? Those moments when life felt perfect, or at least closer to it. You can almost taste the cotton candy at the fair or hear the laughter of friends at a summer barbecue. Nostalgia can be a warm blanket on a cold night, wrapping you in comfort. Yet, it can also be a thief, robbing you of the joy in the present as you dwell too much on what once was.

Regret, on the other hand, is a heavier burden. It's that nagging feeling in your gut, the "what ifs" that keep you up at night. We've all been there—wishing we could rewrite our stories, make different choices, or perhaps take that leap we were too scared to take. Regret has a way of anchoring us in the past, pulling us down like an anchor in a stormy sea. But here's the kicker: while nostalgia and regret can weigh us down, they can also be powerful tools for personal growth.

95

Now, let's dive into how we can harness the power of memory for our own development. First off, reflection is key. Take a moment to sit quietly and think back on your life. What memories stand out? What lessons did you learn from those moments? Maybe it was a failure that taught you resilience, or a success that sparked your passion. Jot these down. You might be surprised at how illuminating this exercise can be.

Next, consider practicing gratitude. Instead of letting regret take the wheel, focus on the lessons you've learned from your past. Gratitude shifts your perspective from what you lack to what you have gained. When you appreciate the journey, you start to see the beauty in every twist and turn.

Another technique is visualization. Picture
yourself stepping into those memories, feeling
the emotions as if you were living them again.
What did you learn? How did it shape who you
are today? This practice not only helps you
understand your past but also empowers you to
shape your future. You're not just a passive
observer; you're the author of your own story.

And don't forget the power of storytelling.
Share your experiences with others. Whether
it's in a journal, a blog, or a casual chat over
coffee, storytelling can be cathartic. It allows
you to process your memories and see them
from a new angle. Plus, you never know who
might resonate with your tale. Your story could
inspire someone else to take a leap or embrace
their own journey.

As you navigate the waters of memory, keep
in mind that your unique perspective is your
greatest asset. No one else has lived your life,

felt your feelings, or learned your lessons. Embrace that individuality. Your memories are not just relics of the past; they are the keys to unlocking your future.

So, let's wrap this up with a little encouragement. Your memories are powerful. They shape your understanding of time, influence your emotions, and hold the potential for growth. Embrace them, learn from them, and let them guide you as you continue your journey. You've got this! Remember, every moment is a stepping stone to the next, and with each memory, you're crafting a richer, fuller life. So go on, unlock those doors and step boldly into your future. The world is waiting for your story.

Chapter 9

Time Travel and Technology

In the grand tapestry of time travel, technology is the needle that stitches our dreams into reality. It's a wild ride, my friend, where the past, present, and future dance together in a dazzling display. So, let's dive into the marvels of technology that shape our understanding of time and open doors to possibilities we once thought were mere fantasies.

First up, let's talk about how advancements in technology are reshaping our perception of time. You know, time ain't just a ticking clock anymore. With the rise of neuroscience and cognitive psychology, researchers are peeling back the layers of how we experience time.

They're using brain imaging techniques to explore how our minds interpret the flow of time. Imagine that! We're starting to understand that time perception can be influenced by a multitude of factors—emotions, memories, and even our environment. It's like discovering a new color in a world that used to be black and white.

For instance, studies have shown that when we're engaged in something we love, time seems to fly. Conversely, during moments of boredom or anxiety, those seconds can stretch out like taffy. This knowledge is powerful. It gives us insight into our own lives and how we can manipulate our experiences to savor the moments that matter. The more we learn about time perception, the more we can harness it to enrich our lives. Think of it as having a remote control for your own temporal experience—fast-forward through the dull bits and rewind to relive the golden moments.

Now, let's pivot to virtual reality. Oh boy, this is where it gets exciting! Virtual reality isn't just a fancy gadget for gamers anymore; it's a groundbreaking tool for exploring time. Picture this: you put on a VR headset and suddenly, you're standing in the middle of a bustling marketplace in ancient Rome or witnessing the signing of the Declaration of Independence. It's immersive, it's educational, and it's downright thrilling!

Virtual reality allows us to step into the shoes of our ancestors, to feel the weight of their decisions and the texture of their daily lives. It breaks down the barriers of time, giving us a front-row seat to history. But it doesn't stop there. Imagine using VR to visualize your future, to explore the paths you could take and the consequences of your choices. It's like having a crystal ball that's powered by technology and creativity. You can experiment, learn, and grow without the fear of making irreversible mistakes.

Now, let's not forget about the future technologies that could one day make time travel a reality. I know, I know—it sounds like something straight out of a sci-fi movie, but hold onto your hats! Scientists are currently exploring concepts like wormholes and time dilation. They're delving into the fabric of space-time, trying to figure out if there's a way to bend it to our will.

Take, for example, the idea of a traversable wormhole. It's a theoretical passage through space-time that could connect distant points in time and space. If we could figure out how to stabilize one of these bad boys, we might just have ourselves a time machine! And then there's time dilation, a phenomenon that occurs when you travel at speeds close to the speed of light. According to Einstein's theory of relativity, time moves slower for those traveling at high speeds compared to those who remain stationary. It's mind-boggling!

Imagine astronauts returning from a long mission in space, having aged only a few years while decades have passed on Earth. That's time travel, my friend, albeit in a different form. These advancements are not just the stuff of dreams; they're rooted in scientific exploration and discovery.

But here's the kicker: as we inch closer to unlocking the secrets of time travel, we must also consider the implications of these technologies. What happens when we can visit the past? What ethical dilemmas will we face? The responsibility that comes with such power is immense. It's a thrilling thought, but it's also a heavy burden.

So, as we navigate this exciting terrain of time travel and technology, remember that every step forward is a chance to reflect on our relationship with time. It's a journey filled with possibilities, where we can learn from the past, enrich our present, and shape our future. Embrace the adventure, keep your mind open, and let your imagination soar. You're not just a spectator in this cosmic play; you're a participant, ready to explore the wonders of time.

Let's celebrate the innovations that make our exploration of time possible. With each advancement, we're not just unlocking doors; we're opening windows to new worlds, new experiences, and new understandings of what it means to be human. So, buckle up and get ready, because the ride into the future of time travel is just getting started!

Chapter 10

The Psychology of Time

Time, huh? It's a funny thing. We can feel it tickin' away, but then there are those moments when it feels like it's just frozen in place. Ever notice how time can feel different depending on what you're doin'? For some folks, a minute drags on like a long, lazy Sunday afternoon, while for others, it zooms by like a rollercoaster. It's kinda wild, really. We're all in this big dance of life, each groovin' to our own beat.

Let's dive a bit deeper into how we perceive time. Our brains are pretty fascinating—they take cues from our experiences, emotions, and even the environment around us. Picture this: you're at a family barbecue, laughing and enjoying good food. Time flies, right? But put us in a dull meeting, and suddenly, every second feels like it's stretching into infinity. It's like time has its own personality, influenced by stuff like our age, mood, and cultural background. Some cultures are all about being on time, while others? Well, they're more laid-back, riding the wave of the moment.

And speaking of time, let's chat about time travel. I mean, who hasn't daydreamed about hopping into a time machine? The idea of zipping through time can get the heart racing. But think about the psychological impact—what if you could step back into your childhood home or jump ahead to see the world a century from now? It's thrilling, but also kinda scary.

Time travel could bring up all sorts of feelings—nostalgia, joy, or even regret. Imagine opening a box of old memories you thought you'd tucked away for good. Those memories can hit you hard, like a freight train outta nowhere. The excitement of exploration can quickly morph into an emotional rollercoaster, taking you for a ride you didn't sign up for.

Now, let's flip the coin and talk about time-related anxiety. Oh boy, that's a heavy topic. The pressure of time can feel like a weight on our shoulders. We stress about missed chances, the ticking clock of aging, and the never-ending deadlines. It can make you feel like you're caught in a tornado, spinning outta control. But hey, there's a silver lining—there are ways to cope with that anxiety.

First off, let's chat about mindfulness. It can be a total game-changer. When you focus on the present, it helps ease that nagging worry about what's coming or what's already happened. It's like taking a deep breath and reminding yourself that, right now, you're okay. Engage in activities that ground you—go for a walk, meditate, or just enjoy a cup of coffee. Those moments of stillness can help you reclaim your sense of time and cut down on anxiety.

Another powerful tool? Perspective. When you find yourself stressing about time, ask yourself: What really matters? What's worth your time and energy? Shifting your focus to what brings you joy and fulfillment can reshape your relationship with time. It's all about prioritizing what fills your cup and letting go of the rest.

And hey, don't forget to celebrate your wins, no matter how small! Every step you take, every moment you cherish, adds up. Recognizing your achievements can help you feel more in control of your time, instead of feeling like it's slipping away.

At the end of the day, the psychology of time is this rich tapestry woven from our experiences, emotions, and cultural influences. Whether we're reflecting on the past, peeking into the future, or just trying to figure out the present, our relationship with time shapes our lives in some pretty profound ways. So as you navigate your own journey through the time warp, remember: you've got the key to how you perceive and manage time. Embrace it, learn from it, and let it lead you to a more fulfilling life.

Time's not your enemy; it's your ally. With a bit of understanding and a sprinkle of mindfulness, you can unlock the secrets of time's doorway and step into a world full of possibilities. So go on, take that leap! The adventure awaits.

Now, let's break this down even further.

Think about how we measure time. Clocks, calendars, and schedules—they're all tools we use to keep track of time. But what if we didn't have them? Would we feel more free or more lost? In some cultures, time isn't measured so rigidly. They flow with the rhythm of life, prioritizing relationships and experiences over schedules. It's a beautiful way to live, but it can also lead to misunderstandings when interacting with more time-bound cultures.

Imagine you're in a business meeting, and everyone's buzzing about deadlines. You're trying to keep up, but your mind drifts to that sunset you saw last night or the laughter of friends at dinner. Suddenly, you feel that familiar twinge of anxiety creeping in. You start thinking about how quickly time flies and how you're not accomplishing enough. But what if, instead, you let that moment of nostalgia wash over you? What if you took a second to breathe, to appreciate that memory, and then brought that energy back to the present?

And let's not forget about the tech we have nowadays. With smartphones, we're always connected, always aware of the time. Notifications buzz, reminders pop up, and suddenly, it feels like time is running away from us. It's a double-edged sword. On one hand, we can manage our time better, but on the other, it can create a sense of urgency that's hard to shake.

Ever catch yourself scrolling through social media, losing track of time? You sit down for "just a minute" and end up hours deep in someone's vacation photos. It's easy to get sucked in, right? That's the trickiness of time perception. What feels like a quick scroll can turn into a time sink. It's like a black hole, pulling you in and making you forget about everything else.

But here's the kicker—what if we flipped that script? What if we used our devices to enhance our time management instead of letting them control us? Setting boundaries, scheduling breaks, or even using apps that remind us to step away from screens can help us reclaim our time.

Now, let's chat about age and how it affects our perception of time. Ever notice how the older you get, the faster time seems to fly? It's like when you're a kid, summer vacation feels like it lasts forever, but as an adult, it's like you blink and it's gone. Some psychologists say it's because we have fewer new experiences as we age. Kids are constantly learning and experiencing new things, making time feel richer and fuller. As adults, we settle into routines, and everything starts to blur together.

So how do we combat that? Embrace new experiences! Try something you've never done before—take a dance class, go hiking, or travel somewhere new. Those fresh experiences can make time feel more expansive. It's like adding color to a black-and-white movie.

And let's not forget about the role of emotions in our perception of time. Ever been so happy that time just slips away? Or so sad that it drags on? Emotions can warp our sense of time. When you're in a great mood, you might find yourself saying, "Wow, where did the time go?" But when you're down, it can feel like time is just crawling.

Here's a thought: what if we leaned into those emotions? Instead of trying to rush through the tough times, what if we sat with them for a bit? Embracing our feelings can help us process them, and in turn, might shift our perception of time. It's like giving yourself

permission to feel and be present, no matter the
situation.

And let's not forget the importance of
community. Our relationships with others can
shape how we experience time. Think about it:
when you're surrounded by loved ones, time
seems to fly. But when you're alone, it can feel
like it drags on. Building strong connections
with people can create those moments of joy
that make time feel richer.

So, how do we cultivate those connections?
It's simple—make time for the people who
matter. Whether it's a phone call, a coffee date,
or a game night, prioritize those moments.
You'll find that the more you invest in your
relationships, the more fulfilling your
experience of time becomes.

Now, let's wrap this up. The psychology of time is complex, but it's also deeply personal. It's shaped by our experiences, emotions, culture, and the choices we make every day. So, as you move through your life, remember that you have the power to shape your relationship with time.

Time can be your friend or your foe—it all depends on how you choose to engage with it. So, go ahead, embrace the moments, learn from the past, and step boldly into the future. Your adventure awaits, and time is just a part of the journey.

And hey, don't forget to enjoy the ride!

Chapter 11

The Future of Time Travel

As we stand on the brink of what seems like a time travel revolution, it's a wild ride, ain't it? Just think about it—predictions are swirling around like leaves in a fall breeze, hinting at advancements that could redefine how we see time itself. Scientists, dreamers, and storytellers are all in the mix, each contributing to a tapestry woven with possibilities. Buckle up, my friend, because the future of time travel is not just a distant dream; it's right around the corner.

Imagine a world where time travel isn't just a concept reserved for sci-fi movies or dusty old books. We're talking about tangible advancements, folks! With quantum mechanics

on our side and technology that's evolving
faster than a kid on a sugar high, the potential
for real-time travel is becoming more than just a
fantasy. Researchers are already toying with
ideas like wormholes and warp drives—fancy
terms, sure, but they represent the spirit of
exploration. What if, one day, we could hop into
a device and zip back to witness the signing of
the Declaration of Independence? Or maybe,
just maybe, we could take a peek into the future,
catching glimpses of what's to come.

Now, let's not kid ourselves; with great
power comes great responsibility. The societal
impacts of time travel could be monumental.
Picture it: the ability to change past events could
reshape our world in ways we can't even
fathom. It's like opening a Pandora's box of
possibilities. Will people use this power for
good? Or will some folks, driven by greed or
revenge, rewrite history to soerve their own
interests? The ethical implications are as vast as
the universe itself, and we've got to tread

carefully. We could see a world where people are more empathetic, learning from their past mistakes, or perhaps a society where time travel is just another tool for manipulation. It's a double-edged sword, and we must be prepared for the consequences.

And here's where science fiction struts in, wearing a cape and a grin. It's been shaping our understanding of time travel since the days of H.G. Wells. Those fantastical tales, with their imaginative gadgets and paradoxical plots, inspire real-life scientists to push the boundaries of what's possible. They spark our curiosity and fuel our desire to explore the unknown. Think of the classic films and books that have captured our hearts—each story a stepping stone, guiding us toward a future where time travel might just be a reality. Science fiction isn't just entertainment; it's a blueprint for innovation, a call to action for dreamers and doers alike.

Let's break this down a bit, shall we? First off, predictions for time travel advancements are all about innovation. We're in an age where technology is growing faster than a field of wildflowers in spring. Researchers are diving deep into the realms of theoretical physics, exploring ideas that were once considered pure fantasy. Theoretical devices like time machines are no longer just figments of our imagination; they're becoming serious topics of discussion. We're talking about harnessing energy in ways that could allow us to bend the fabric of time itself. It's exhilarating to think about the breakthroughs that could be just around the corner.

Now, consider the societal impacts. The ability to travel through time could lead to unprecedented changes in our culture, our politics, and even our personal lives. What if we could go back and prevent wars, heal rifts between nations, or even just mend broken relationships? The potential for positive change is staggering. But let's not forget the darker side of this coin. The power to alter time could lead to chaos, with people trying to erase their mistakes or rewrite their narratives to fit a more favorable light. This could create a society where the past is malleable, and trust becomes a rare commodity. We need to think long and hard about the moral responsibilities that come with such capabilities.

And then there's the role of science fiction. It's like a guiding star, illuminating the path ahead. Those stories we grew up with—whether it was "Back to the Future" or "Doctor Who"—have ignited our imaginations and inspired generations of thinkers. They ask the big questions: What would happen if we could change the past? How would our present be affected? Science fiction doesn't just entertain; it provokes thought and inspires innovation. It pushes the boundaries of what we believe is possible, nudging scientists to explore ideas that may seem outlandish today but could be tomorrow's reality.

So, what's the takeaway here? The future of time travel is a tapestry of hope and caution. It's a dance between possibility and responsibility. As we stand on the cusp of these advancements, let's embrace the journey ahead. Imagine the stories we'll tell, the lives we'll touch, and the lessons we'll learn along the way. It's not just about the destination; it's about the adventure, the growth, and the connections we make through time.

Now, take a moment to visualize your own role in this unfolding narrative. What do you want to contribute to this future? Your voice, your ideas, and your passion can shape the world of tomorrow. So dream big, stay curious, and remember: the future is bright, and it's waiting for you to step into it. You've got this!

Chapter 12

Interpersonal Relationships Across Time

Life's a winding road, isn't it? We stumble and fumble our way through, often getting tangled in the web of relationships we build along the way. But what if I told you time travel isn't just about fancy machines and zipping through eras? Nope, it's also about connecting with those past versions of ourselves and the folks who've drifted in and out of our lives. So, let's dig deep into this topic and see how we can unlock the secrets of our interpersonal relationships across time.

First up, let's chat about our past selves. Imagine standing in front of a mirror, but instead of your reflection, you see a younger

you—maybe it's that awkward teen, the eager
college kid, or the version of you who was just
trying to figure it all out. Those past selves are
like pieces of a puzzle, each holding a story
that's shaped who you are today.

Engaging with these past versions can really
change the game. It's like catching up with an
old buddy. You get to reflect on choices you
made, lessons you learned, and yeah, even the
mistakes you'd love to erase. But here's the
twist: instead of pushing those mistakes away,
embrace them. They're not just scars; they're
badges of honor that tell the tale of your growth.

So, when you look back at your past self,
what would you say? Maybe it's to take that
leap of faith or to hold onto friendships that felt
so fleeting. By nurturing that relationship with
your past self, you can really dig into
understanding who you are now. It's like

weaving together the fabric of your identity, one thread at a time.

Now, let's switch gears and talk about how time travel messes with love and friendship. Ah, love—the most timeless emotion of all. When we think about relationships, it's easy to get swept up in the chaos of the present. But what about those connections that have faded? You know, the friendships that once brought joy, laughter, and maybe a sprinkle of heartache?

Time's got a funny way of changing how we see things. It can deepen love or create distance, depending on how we handle it. Think of love like a river, flowing through time. Sometimes it's a gentle stream, other times a raging current. The trick is learning to ride those waves.

Imagine reconnecting with an old friend you lost touch with ages ago. You've both traveled

through different seasons of life, but that bond?
It's still there, just waiting to be reignited.
Reconnecting isn't just about nostalgia; it's
about recognizing the growth you've both gone
through. It's like stumbling upon a treasure
chest filled with memories, laughter, and maybe
a few tears.

And love—whether it's romantic or
platonic—has its own beauty. Time travel lets
us revisit those pivotal moments in our
relationships. What if you could go back to that
first date, the moment you knew you were in
love? Or maybe it's that falling out with a friend
where you wish you could rewrite the ending.

Here's a thought: why not create a mental
time capsule? Fill it with memories of your
relationships, the highs and the lows. When you
dig into that capsule, you'll see how far you've
come and how those experiences shaped your
understanding of love and friendship. It's a

beautiful way to honor the past while embracing the present.

Let's dive into the art of reconnecting with lost connections through time. Life pulls us in all sorts of directions, and sometimes we lose touch with those who once meant everything. But guess what? Time travel isn't just a fantasy; it's a mindset. You can reach out to those connections, even if it's been ages.

Think about it: have you ever scrolled through old messages or social media posts and found a convo that made you smile? That's your sign! Send a message, or just pick up the phone. You'd be surprised how a simple "Hey, how's it going?" can open the door to rekindling a friendship or even sparking a romance.

But here's the catch—be ready for the ride. Just like time travel can be a wild ride, reconnecting can be unpredictable too. People change, life happens, and sometimes the connection you once had might not fit anymore. And that's totally okay! Embrace the journey for what it is, and be open to new possibilities.

Now, let's take a moment to reflect on how these relationships across time impact our lives. Each connection, each moment, shapes us into who we are today. They teach us resilience, compassion, and the importance of being present. So, as you navigate this intricate web of relationships, remember to cherish the past, embrace the present, and look forward to the future.

Picture your journey through time. Imagine standing at the crossroads of your relationships, looking back at the paths you've traveled. Each relationship is a chapter in your story, filled with lessons and memories.

Whether you're reconnecting with a past self, riding the waves of love and friendship, or reaching out to lost connections, remember this: you hold the key to unlocking the secrets of your interpersonal relationships across time. Embrace it, cherish it, and let it guide you on this incredible journey of self-discovery and connection.

You've got this! Your relationships, both past and present, are waiting for you to dive in and explore. So go ahead, take that leap, and see where the journey leads you. The door to your temporal relationships is wide open—step through and make the most of it!

Now, let's dig a bit deeper. Relationships aren't just about the people we meet; they're also about how we evolve. Think about the friendships that faded away. Some people come

into our lives for a reason, a season, or a
lifetime. It's a classic saying, but it rings true.

Ever had a friend who was your ride-or-die
for a few years, only to drift apart? Life
happens. Maybe one of you moved, or you took
different paths. It's like a chapter ending in a
book. But just because that chapter closed
doesn't mean the story's over.

And what about those relationships that left
a mark? You know the ones—those friends who
taught you something valuable, even if it was
through heartbreak. They shaped your
understanding of love, trust, and sometimes
even betrayal.

Reflecting on these relationships can be
tough, but it's also healing. It's like cleaning out
your closet. You find old clothes that don't fit
anymore, but they remind you of who you were.

Letting go of what doesn't serve you is just as important as holding onto what does.

Now, let's talk about forgiveness. Forgiving yourself for past mistakes is crucial. We all mess up. Maybe you ghosted a friend or said something hurtful. It happens. But holding onto that guilt? It weighs you down. So, cut yourself some slack. Learn from it, and move on.

And what about forgiveness towards others? It's not always easy. Sometimes, it feels like you're carrying a heavy backpack filled with rocks—each one representing a grudge or a hurt. But when you let go of that weight, it's like shedding layers of an old skin. You feel lighter, freer.

Think of it this way: relationships are like a garden. You've got to tend to them. Some plants thrive, while others wither. You've got to pull out the weeds—those toxic connections that drain your energy. And then, nurture the ones that bring you joy.

Let's not forget about the role of communication. It's the lifeblood of any relationship. Without it, things can get messy. Ever had a misunderstanding because someone didn't speak up? Yeah, it happens. Being open and honest can save a friendship from crumbling.

And while we're on the topic, don't underestimate the power of vulnerability. It's scary to open up, but it's also incredibly rewarding. When you share your fears and dreams, you invite others to do the same. That's how deep connections are formed.

Now, let's get a little philosophical. What if time isn't just linear? What if it's more like a spiral, where past, present, and future intertwine? Think about how your past experiences influence your current relationships. They shape your expectations, your fears, and your desires.

Ever had a situation where you reacted strongly to something because it reminded you of a past hurt? That's the past creeping into the present. Recognizing these patterns can help you break free from them. It's like catching a glimpse of a shadow and realizing it doesn't define you.

As we explore these themes, let's not forget about the beauty of new beginnings. Just because a chapter ends doesn't mean the book is closed. New friendships can blossom at any stage of life. Whether you're starting a new job, moving to a new city, or just stepping out of your comfort zone, there are opportunities everywhere.

Picture this: you're at a coffee shop, minding your own business, and someone strikes up a conversation. Before you know it, you're sharing stories and laughter. That's the magic of connection. It can happen anywhere, anytime.

But here's the thing: don't be afraid to put yourself out there. Take that leap of faith. Join a club, take a class, or volunteer. You never know who you'll meet or what connections you'll make.

And as you navigate these new relationships, remember to be patient. Building trust takes time. It's like planting a seed and waiting for it to grow. You've got to water it, nurture it, and give it the right conditions to thrive.

So, as we wrap this up, let's take a moment to appreciate the journey. Your relationships, both past and present, are like a tapestry woven together with threads of experience, love, and growth. Embrace the messiness, the beauty, and the lessons that come with it.

In the end, it's all about connection. It's about reaching out, being vulnerable, and celebrating the moments that shape us. So, go ahead—dive into those memories, reconnect with your past self, and cherish the relationships that mean the most to you.

You've got the power to shape your narrative. Your journey through time is uniquely yours, filled with twists and turns that lead to growth and discovery. So step through that door and see where it takes you. The adventure of a lifetime awaits!

Chapter 13

Cultural Interpretations of Time

Time, my friend, is a fascinating beast. It flows differently depending on where you stand on this big ol' globe. In some cultures, it's a river—always moving, always changing. In others, it's a circle, looping back on itself, where the past, present, and future are intertwined in a dance as old as time itself. Each perspective offers a unique lens through which we can explore the concept of time travel, revealing not just how we perceive time but how those perceptions shape our narratives about hopping through it.

Let's take a stroll through a few cultures, shall we? In many Western societies, time is treated like a commodity. It's all about schedules, deadlines, and the relentless ticking of the clock. You know the type—"Time is money," they say. This perspective can create a sense of urgency, pushing folks to chase after moments, often leading to a feeling of scarcity. The clock rules, and every tick is a reminder of what's slipping away. It's a fast-paced world, where the future is always just out of reach, and the past? Well, it's often seen as a place of regret.

Now, flip the coin to some Eastern philosophies, where time is more cyclical. Think about the rhythms of nature, the changing seasons, and the way life ebbs and flows. In cultures like those influenced by Buddhism, time is viewed as an endless cycle of birth, death, and rebirth. Here, the past is not just something to learn from but a vital part of the present, shaping who we are. The future? It's an open canvas, influenced by the choices we make in this very moment. This view can be liberating, allowing for a sense of peace and acceptance as we navigate our journeys.

And what about indigenous cultures? They often see time as a web, intricately woven with stories, ancestors, and the land itself. For them, the past is alive, breathing through rituals and traditions that connect generations. Time travel, in this sense, isn't just a fantasy—it's a lived experience. They understand that every action has a ripple effect, echoing through time, shaping their reality and that of future generations. It's a profound reminder that we are all part of something larger, a continuum of existence.

Now, let's dive into how these cultural attitudes impact our narratives around time travel. When we think about time travel in literature and film, we often see these cultural perspectives reflected in the stories we tell. For instance, in Western narratives, time travel is frequently portrayed as a way to escape the present or correct past mistakes. Movies like "Back to the Future" and "The Time Machine" embody this idea of racing against the clock, trying to fix what's gone wrong. It's thrilling, no doubt, but it also reflects that urgent, often frantic approach to time.

On the flip side, consider narratives from cultures that embrace cyclical time. Stories like "Groundhog Day" highlight the idea of learning from the past, suggesting that time travel isn't just about changing events but about personal growth and understanding. Here, the protagonist is given a chance to relive moments, to make different choices, and ultimately to discover the beauty in the present. It's a refreshing take, one that encourages us to reflect on our actions and their impact on our lives and the lives of others.

And let's not forget the power of indigenous storytelling, where time travel often transcends the physical. In many tales, characters journey through time not just to change events but to connect with their ancestors, to understand their roots, and to honor the wisdom of those who came before. These narratives remind us that time is not just a linear path but a rich tapestry of experiences that shape our identity.

As we celebrate these diverse temporal philosophies, it's essential to recognize the beauty in our differences. Each cultural interpretation offers us a unique insight into the human experience. They teach us that time is not just a measurement but a profound aspect of our existence, shaping our beliefs, our stories, and ultimately, our lives.

So, how do we embrace this kaleidoscope of perspectives? Start by reflecting on your own relationship with time. How do your cultural background and personal experiences shape the way you view time? Are you more of a clock-watcher, or do you find comfort in the cyclical nature of life? Take a moment to ponder that.

Next, let's weave these insights into your writing. When crafting your time travel narrative, consider how cultural attitudes can influence your characters' motivations and actions. Perhaps your protagonist hails from a culture that sees time as a circle, leading them to approach their journey with a sense of reverence and respect. Or maybe they're driven by a need to correct past mistakes, reflecting that urgent, linear perspective.

And remember, you don't have to stick to just one interpretation. Mix and match! Create characters from diverse backgrounds, each bringing their unique understanding of time to the table. This not only enriches your narrative but also invites readers to explore the complexities of time through different lenses.

Lastly, don't shy away from the philosophical implications of time travel. Use your narrative to challenge readers' perceptions. What if time isn't just a series of moments to be navigated? What if it's a vast ocean, full of depths to explore and mysteries to uncover? Encourage your readers to think deeply about their relationship with time, and how it shapes their lives.

As you embark on this writing journey, keep in mind that your unique voice and perspective are your greatest assets. Don't be afraid to share your own experiences and reflections on time. Personal stories create powerful connections with readers, allowing them to see themselves in your narrative.

In the end, time is more than just a backdrop for your story—it's a character in its own right.

Embrace its complexities, celebrate its diversity, and let it guide your narrative in ways that inspire and transform. You've got this! So grab your pen, let your imagination soar, and unlock the secrets of time travel's doorway. The world is waiting for your unique perspective, and trust me, it's gonna be a wild ride!

Chapter 14

The Paradoxes of Time Travel

Time travel's a wild ride, right? It's not just about hopping in a DeLorean and zooming back to the past like you're on some cosmic joyride. Nah, it's way more complicated than that. Think of it as a crazy puzzle, full of twists and turns that can leave your brain doing somersaults. Ever heard of the grandfather paradox? It's a classic. You pop back in time and accidentally stop your grandparents from ever meeting. If that happens, how in the heck do you even exist to travel back in time? It's like trying to untangle a ball of yarn that just keeps knotting itself up the more you mess with it.

These time travel paradoxes? They're not just brain teasers; they mess with how we see cause and effect. They make us question everything we thought we knew about reality. But hey, don't let that freak you out! Think of it as a nudge to dig deeper into what time, existence, and our role in the universe really mean.

Let's break it down a bit. Imagine you hop into a time machine, all set to witness some epic historical moment. But wait! What if your very presence changes the course of history? You might think you're just there to watch, but even the tiniest thing you do could send shockwaves through time, kinda like tossing a pebble into a still pond. Every choice you make creates a whole new timeline, branching off into who knows where.

When you're faced with these timeline conflicts, try to keep an open mind. Instead of seeing time as this straight, narrow road, picture it as a massive landscape with tons of trails.

Some paths might lead to dead ends, while others could open up whole new worlds. It's all about finding a groove in the chaos.

So, how do we dodge these pesky paradoxes in the first place? There's a ton of theories floating around, from the totally wacky to the scientifically grounded. One popular idea is "self-consistency." This one suggests that any move a time traveler makes was always part of the story. If you go back to prevent a disaster, guess what? Your attempts might just end up causing that disaster in the first place. It's like trying to catch smoke with your hands— impossible and frustrating.

Then there's the multiverse hypothesis. Picture this: every choice you make creates a new universe, branching off into endless possibilities. If you mess with the past in one universe, you're just opening the door to a new timeline where everything plays out differently.

This way, you can explore all those "what ifs" without getting bogged down by paradoxes.

But here's the kicker: these paradoxes and theories aren't just nerdy puzzles for scientists to chew on. They hit close to home. Think about it—how often do we find ourselves stuck in a loop of decisions, wishing we could change the past while stressing about the future? Understanding time travel paradoxes can actually give us some pretty deep insights into our own lives.

As you wade through the complexities of time travel, keep in mind to embrace the unknown. Each paradox is a lesson, a chance to grow and learn. Whether you're contemplating the impact of your choices or just trying to figure out your own life's timeline, stay open-hearted. The journey through time—whether it's literal or metaphorical—can be a real game-changer.

And hey, if you ever feel overwhelmed, just take a breather. The beauty of time travel, whether in books or in our everyday lives, is that it reminds us every moment counts. Each tick of the clock is a chance to learn, to change, and to dive headfirst into the chaos. You got this!

Now, let's dig a little deeper into these time travel paradoxes. They're not just for sci-fi flicks; they can actually help us understand our lives better. For instance, take the butterfly effect. It's that idea that a tiny change in the past can lead to massive differences in the future. You know, like if a butterfly flaps its wings in Brazil, it could cause a tornado in Texas.

This concept makes you think about how every little decision matters. It's like when you're deciding whether to go to that party or stay home. You might think it's no big deal, but that one choice could lead to meeting someone who changes your life forever—or not. The stakes feel high, right?

And then there's the idea of predestination. It's a mind-bender. If everything's already set in stone, then what's the point of making choices? Are we just puppets in some cosmic play? Or is there room for free will? This question can really mess with your head, but it's also super fascinating.

Think about it: if you knew you were destined to become a famous musician, would you still put in the effort to practice? Or would you just coast along, thinking it's all gonna happen anyway? That's where the paradox kicks in. The idea of fate versus free will is a classic dilemma, and it's been explored in

countless stories, from Greek mythology to
modern novels.

Now, let's chat about alternate realities. The
multiverse theory suggests there are countless
versions of you out there, each living different
lives based on the choices you make. Ever
wonder what your life would look like if you'd
taken that job offer in another city? Or what if
you'd pursued a different passion? It's like
having a whole library of "what could've been"
stories, each with its own twists and turns.

This idea can be comforting. It means that
even if you mess up, there's a universe where
you nailed it. But it can also be a bit
overwhelming. How do you choose the right
path when there are so many options? It's like
standing in front of an all-you-can-eat buffet
and not knowing what to pick.

And speaking of choices, let's not forget about the consequences. Every action has a reaction, right? This principle is at the heart of many time travel stories. When characters mess with time, they often face unexpected fallout. It's a reminder that our decisions, big or small, can have ripple effects we might not even realize.

Imagine you're in a movie, and you decide to go back and tell your younger self to avoid that bad relationship. Sounds good, right? But what if that relationship was the reason you found your true passion later on? It's a classic case of "be careful what you wish for."

In our everyday lives, we often grapple with similar dilemmas. Should you take that job that pays more but requires you to move away from friends? Or should you stay put and chase your dreams? These choices can feel monumental,

and they often come with a hefty dose of
uncertainty.

Now, let's take a step back and look at how
these time travel concepts play out in our
personal narratives. Life's kinda like a time
machine, don't you think? We're constantly
traveling through our own timelines, making
choices that shape who we are. Sometimes we
look back and think, "Man, I wish I could
change that." But what if those "mistakes" are
what led us to where we are now?

The beauty of life lies in its unpredictability.
It's messy, chaotic, and full of surprises. Just
like time travel. So, the next time you find
yourself stuck in a loop of regret or anxiety
about the future, remember: you're not alone.
We're all just trying to navigate this wild ride
together.

And let's not forget about the lessons we learn along the way. Every paradox, every conflict, every choice—these are all part of the journey. They teach us resilience, adaptability, and the importance of living in the moment.

So, what's the takeaway here? Time travel may be a fictional concept, but the ideas it brings up are very real. They mirror our own experiences, struggles, and triumphs. Whether you're pondering the butterfly effect or wrestling with the idea of fate, remember that each moment is a chance to grow.

In the end, it's about embracing the chaos and uncertainty of life. Just like time travel, it's unpredictable and wild. But that's what makes it beautiful. So, keep your heart open, take risks, and don't be afraid to explore the unknown.

And when life gets overwhelming, just breathe. Every tick of the clock is a reminder that you're alive, and that's something to celebrate. So go out there and make the most of your journey—time travel or not. You got this!

Chapter 15

Embracing the Journey

Life's a crazy ride, isn't it? And time—man, time's the bus we're all stuck on. When we dig into what time really is, it's clear it ain't just a straight line from yesterday to tomorrow. Nope, it's more like a colorful quilt stitched together with our experiences, memories, and those little moments that make us who we are. Finding meaning in this time trip is like uncovering hidden gems in your own backyard. You just gotta dig a bit deeper.

Think about it. Every tick of that clock? It's an opportunity. A chance to learn something new. When we step back and look at our journey through time, we start seeing how everything connects—our past, present, and future. Each moment is like a thread in our life's fabric, and every experience—good, bad, or a total train wreck—adds its own splash of color. Embracing this journey means realizing that every twist and turn has its worth. It's not just about where we end up; it's about enjoying the ride along the way.

Now, let's chat about living in the moment. Ah, the present—what a slippery little rascal! We often get tangled up in the past, rehashing old memories, or stressing over what's coming next. But here's the kicker: the present is where all the magic happens. It's the only moment we've really got, and it's bursting with possibilities. When you tune into the now, you open yourself up to experiences that can flip your world upside down in the best way possible.

Picture this: you're chilling at your go-to coffee spot, sun streaming through the window, sipping on that perfect cup of joe. Instead of scrolling through your phone or stressing about tomorrow's endless to-do list, you take a deep breath and really soak in the moment. You hear the laughter of friends at the next table, watch the barista whip up some latte art like a pro, and feel the warmth of the sun on your skin. That's living in the present, my friend. It's a grounding practice that connects you to everything around you and lets you appreciate the beauty of life as it happens.

But let's not overlook how time travel can be a killer metaphor for our lives. Imagine your life as a time machine, where every choice you make sends ripples through your timeline. Just like in those sci-fi flicks, you can't rewind and change the past, but you can learn from it. Use those lessons to shape your future. The choices you make today? They'll echo into tomorrow, so why not make 'em count?

Think of your journey as a series of timelines. You've got the path you're on, the one you might've taken if you'd made different calls, and the endless possibilities ahead. Embracing the journey means recognizing you're the captain of your ship, navigating the waters of time. You've got the power to chart your course, and every decision is a stepping stone toward your destination.

Let's take a sec to appreciate this metaphor. Just like time travel, life's full of unexpected twists and turns. You might find yourself in a situation you never saw coming, but that's where the growth happens. It's in those moments of uncertainty that you discover your strength, resilience, and creativity. You learn to adapt, pivot, and embrace change. And that, my friend, is where the real magic lies.

Now, I can hear you thinking, "But what if I make the wrong choice? What if I screw up?" Here's the deal: there's no such thing as a wrong choice. Every decision is a chance to learn and grow. Even the flops can lead you to places you never dreamed of. It's all part of the journey, and embracing it means letting go of the fear of failure.

So how do you embrace this wild ride through time? First off, practice mindfulness. Slow down and take a moment to appreciate where you are right now. Breathe in the present and shake off the distractions that pull you away from it. Whether it's through meditation, journaling, or just taking a stroll in nature, find what helps you connect with the here and now.

Next, reflect on your past. What lessons have you picked up? What experiences shaped you into who you are today? Write 'em down. Celebrate your growth. Each experience, whether joyful or painful, has added to your unique story. Own it. Embrace it.

And finally, visualize your future. What do you wanna create? What dreams do you wanna chase? Picture yourself achieving those goals, and let that vision steer your actions. Remember, you're not just a passenger on this journey; you're the driver. The choices you make today will pave the way for the life you wanna live tomorrow.

As you dive into this journey, keep in mind it's not about being perfect. It's about making progress. Celebrate those little wins along the way. Did you take a moment to enjoy a beautiful sunset? Did you reach out to a friend you hadn't spoken to in ages? Did you take a leap of faith and try something new? Each of these moments is a victory worth celebrating.

Embracing the journey means finding joy in the process. It's about being open to the unexpected and allowing yourself to grow. So, next time you feel overwhelmed by the passage of time, remember this: you have the power to shape your experience. You're not just a bystander; you're an active player in your life story.

Let's take a moment to appreciate the beauty of time. It's a gift, a precious resource that lets us explore, learn, and grow. Embrace the journey, my friend. Find meaning in exploring time, live fully in the present, and let the metaphor of time travel guide you on your path. You've got this, and the world's waiting for your unique story to unfold. So buckle up, enjoy the ride, and remember: every moment is a chance to create something extraordinary.

But wait, there's more! Let's dig deeper into this whole concept of embracing the journey.

It's not just about living in the moment or learning from the past; it's about how we interact with time itself. Think about how we often rush through life, always looking for the next big thing. We're so focused on the future that we forget to appreciate what's right in front of us.

Imagine you're at a concert. You're there, the music's pumping, the crowd's alive, but instead of soaking it all in, you're busy snapping pics for Instagram. You miss the vibe, the energy, the sheer joy of the moment. You gotta ask yourself—what's the point? Life's not just about capturing moments; it's about living them. So, next time you find yourself at a concert, put the phone down and just dance. Feel the music.

And let's not forget the people we meet along the way. Each person we encounter adds a new layer to our journey. Think of your best friend. You've shared laughter, tears, and countless memories. Each moment spent together has shaped your story. Embrace those relationships. Nurture them. They're part of your journey, and they make the ride so much richer.

Now, let's talk about setbacks. We all hit bumps in the road, right? Maybe you didn't get that promotion you were gunning for, or a relationship didn't pan out like you hoped. It's easy to get down on yourself in those moments. But what if, instead of seeing them as failures, you viewed them as stepping stones? Every setback is a lesson, a chance to reassess and redirect your course. It's like when you're driving and miss your exit. You don't just sit there, sulking; you find another way to get where you wanna go.

And hey, let's not forget about gratitude. Practicing gratitude can totally change your perspective on time and your journey. When you take a moment to appreciate what you have—your health, your friends, your experiences—you shift your focus from what's missing to what's present. It's like putting on a new pair of glasses that lets you see the beauty all around you.

Ever tried keeping a gratitude journal? It's simple. Just jot down a few things you're thankful for each day. Over time, you'll notice a shift in how you view your life. You'll start to see the abundance instead of the lack. And that, my friend, makes the journey so much sweeter.

Let's not skip over the power of dreams, either. Dreaming big is crucial. It's what keeps us moving forward. But here's the catch: dreams don't just happen; you gotta put in the work. Think of your dreams as a roadmap. You can't just sit back and wait for them to come true. You've gotta take action, step by step, to make them a reality.

Imagine you want to start a business. You've got this great idea, but if you don't take the first step—like writing a business plan or doing market research—nothing's gonna happen. Embrace the journey of pursuing your dreams. Break it down into manageable steps and celebrate each one.

And let's wrap this up with a little humor. Life can be a total circus sometimes, right? You're juggling responsibilities, trying to keep everything balanced, and then boom! A clown shows up (metaphorically speaking, of course) and throws a pie in your face. It's messy, it's chaotic, but it's also hilarious. Embrace the absurdity. Laugh at the chaos. It's all part of the ride.

So, as you navigate through this wild journey called life, remember to embrace every twist and turn. Live in the moment, learn from the past, and dream big for the future. Time's a precious gift, and you've got the power to shape your story. You're not just along for the ride; you're driving this thing.

Take a deep breath, buckle up, and enjoy the ride. Life's too short to be anything but extraordinary. So go out there and make your mark, one moment at a time. You've got this!

Index:

advancements, 12, 13, 37, 99, 103, 116, 119, 121

anxiety, 13, 100, 107, 108, 110, 151

butterfly, 6, 12, 36, 45, 50, 74, 75, 79, 147, 152

challenges, 26, 34, 66, 92

chances, 6, 9, 12, 82, 84, 85, 88, 90, 107

conflicts, 13, 144

connections, 13, 25, 28, 92, 114, 121, 124, 126, 128, 131, 133, 141

Coping, 13

culture, 5, 6, 20, 50, 53, 69, 72, 115, 119, 140

devices, 5, 11, 55, 58, 61, 112, 119

dilemmas, 12, 37, 40, 50, 74, 75, 103, 150

dimension, 11, 18, 22

effects, 13, 70, 75, 150

Energy, 11, 59

fiction, 3, 5, 13, 118, 120

figures, 11, 47, 48, 51

films, 6, 11, 69, 118

foundations, 5, 11, 35

friendship, 9, 13, 124, 125, 126, 128, 131

games, 6, 12, 69, 71

Growth, 12, 80

imagination, 5, 6, 10, 35, 41, 48, 53, 59, 61, 65, 68, 74, 104, 119, 142

impact, 12, 13, 27, 39, 77, 79, 106, 127, 137, 138, 146

innovations, 104

Journey, 14, 154

Learning, 12, 66, 81

literature, 6, 11, 43, 45, 53, 69, 71, 137

love, 13, 44, 51, 100, 123, 124, 125, 128, 129, 134

Memory, 9, 12, 94

metaphor, 5, 12, 14, 48, 82, 156, 157, 160

multiverse, 11, 32, 33, 145, 149

myths, 5, 7, 11, 43, 47, 48, 53

narratives, 5, 6, 13, 120, 135, 137, 138, 139, 151

nature, 6, 9, 16, 23, 38, 44, 136, 140, 158

Navigation, 11, 55

nostalgia, 12, 94, 95, 106, 110, 125

paradoxes, 6, 10, 11, 13, 38, 56, 58, 60, 63, 143, 145, 146, 147 past, 5, 8, 9, 12, 13, 15, 16, 23, 36, 37, 39, 40, 46, 47, 48, 51, 53, 55, 57, 60, 62, 63, 70, 71, 72, 74, 75, 76, 78, 80, 81, 82, 83, 84, 85, 87, 90, 93, 94, 95, 96, 97, 98, 99, 103, 104, 109, 115, 117, 120, 122, 123, 126, 127, 128, 130, 132, 134, 135, 136, 137, 138, 140, 143, 145, 146, 147, 155, 156, 158, 161, 165

perception, 6, 11, 12, 16, 17, 19, 21, 22, 69, 99, 100, 111, 112, 113

perceptions, 5, 6, 9, 11, 19, 24, 25, 135, 141

philosophies, 7, 13, 136, 139

physics, 6, 7, 11, 18, 21, 31, 32, 41, 45, 52, 56, 62, 119

possibilities, 6, 8, 10, 13, 18, 22, 32, 35, 39, 41, 45, 48, 53, 57, 63, 64, 70, 99, 104, 109, 116, 117, 127, 145, 155, 157

Potential, 13

Predictions, 13

present, 6, 8, 10, 14, 15, 16, 20, 23, 39, 46, 48, 53,

60, 78, 82, 95, 99, 104, 107, 109, 111, 114, 120, 124, 126, 127, 128, 132, 134, 135, 136, 138, 155, 156, 158, 160, 163

Quantum, 11, 32, 52

realities, 11, 55, 57, 58, 60, 149

reality, 5, 6, 8, 9, 12, 19, 22, 32, 34, 39, 51, 52, 62, 63, 64, 65, 99, 101, 102, 118, 121, 137, 143, 164

reflection, 20, 53, 87, 96, 122

regret, 12, 94, 95, 96, 106, 136, 151

Relationships, 13, 122, 128

relativity, 11, 17, 22, 31, 33, 34, 35, 45, 49, 103

Responsibilities, 12

societies, 13, 135

storytelling, 87, 97, 138

Techniques, 12

Technology, 12, 99

Theories, 11, 13

TV, 39, 50

www.ingramcontent.com/pod-product-compliance
Lightning Source LLC
Chambersburg PA
CBHW061453150726
47987CB00001B/434